BANANAS PHYSICS

(All You Wanted to Know About Quantum Mechanics

But Were Too Ashamed to Ask)

by

Xavier Perez-Pons

ISBN: 9798639095580

To Laia, Roger, Pau and Júlia

My nephews

With all my love

WHAT EXACTLY IS THE STANDARD MODEL IN THE MOST BASIC TERMS?

The Standard Model is a theory that roughly describes what happens in the most intimate core of everything. Take for example this pen. At first glance, this pen seems like a lifeless thing, where nothing is happening, where everything is calm and without any incident. And yet, in the innermost part of this pen, a drama is taking place compared with which the most uproarious television series and soap operas seem boring documentary films where the time passes uneventfully and one even would say that the TV screen is off.

WHAT IS A NEUTRON STAR?

Without any doubt, the neutron is the star of subatomic particles. Currently, in my MIT laboratory, we have 15 neutrons and two neutrinos. A real luxury. At Columbia University we didn't have access to neutrons. We had to experiment with dust particles and then extrapolate the conclusions. To explain what a neutron is, the best way is with the example of the drunkard who suddenly starts singing in the pub and everyone tries to quieten him. If quiet is achieved, a chain reaction occurs and the crockery gets fired out the window. Those who try to quieten the neutron are the protons, also called "quieteners" or "punchers". However, the neutron resists being punched since it is much stronger than the proton. As a general rule, at least twenty protons are needed to subdue a neutron. Even so, sometimes the neutron emerge victorious, thus forcing the protons to change tactics. They camouflage themselves by pretending to be Egyptian statues. In that case, there is no chain reaction and you have to wait until the next new moon to repeat the experiment.

IN THE GROUND STATE OF THE HYDROGEN ATOM, THE ELECTRON HAS A TOTAL ENERGY OF -13.6 eV. WHAT ARE (a) ITS KINETIC ENERGY AND (b) ITS POTENTIAL ENERGY IF THE ELECTRON IS ONE BOHR RADIUS FROM THE CENTRAL NUCLEUS?

I absolutely refuse to answer such a question. I'm bound to professional secrecy. Not even under torture could you get the answer out of me! Understand this: particle physicists have many flaws, but betraying secrets is not one of them. Enough is enough! What do you take me for? I'll protect the secret with my very life. The code of ethics is clear on this point. If I answered your question, I would not only be betraying the code but I would be jeopardizing the relative stability of subatomic particles. In no way I'll do such thing unless I am offered a good sum of money.

WHAT'S NEUTRON DOING IN THE NUCLEUS?

You don't want to find out, trust me. I've been peeping in the atomic nucleus for longer, and I've seen things that would make Oktoberfest seem dull by comparison. But it's all the fault of the up and down quarks. Neutrons are good guys. They're teetotallers and you won't see them cheating on their wives with other women. Up and down quarks are the ones who have fallen into bad company, such as strange quarks and charm quarks, which just go around drunk, tearing' up stuff and flirtin' with antiquarks. Of course, this has an impact on neutrons' stability. But that's inevitable. It is as if you carry an Oktoberfest within you. You would suffer hallucinations whether you want to or not, and before you knew it, you would be dancing naked and wearing a lamp for a hat.

WHAT IS THE PROOF THAT NEUTRONS EXIST?

Okay, you are dummies, I get the picture. But please refrain any offensive questions. Do you see these scars? Do you see this arm in a sling? Do you see this shiner? Do you see this wheelchair? They are not the result of parachute jumping without a parachute, you know? I have been dealing with neutrons since I was a child. My father organized neutron parties at home. He was a distinguished member of the Philadelphia Neutron Fan Club. I know neutrons like the back of my hand. Every day I experiment with them in my MIT lab. Have you ever heard of MIT at the very least? With all this background, my word should suffice. You are lucky that I am not prone to violence, because otherwise right now I would get up from this wheelchair and kick your teeth in.

WHAT IS A 'BANANA-EQUIVALENT DOSE' IN RADIATION?

BED or 'Banana Equivalent Dose' is a handy way of measuring the quantity of radioactive isotopes in whatever. For example: in bananas. The radiation in a banana is 1 BED. The radiation in 2 bananas is 2 BED. The radiation in 3 bananas is 3 BED. And following this witty procedure one can go up to 2500 BED, which is the maximum permitted radiation in a nuclear power plant. So you better not eat more than 2500 bananas at once. By the way, here are two cases of excessive intake of bananas... During the 1929 stock-market crash, most suicidal stockbrokers killed themselves by jumping from the top of a skyscraper, which denote a pretty lack of imagination. But there was a fanciful broker who traded in bananas, had a large stock of bananas, and knew bananas' radioactive properties. Also, he liked bananas. So he locked himself in his bananas' Brooklyn warehouse and started eating bananas nonstop. Around the same time, there was a Chicago thug who was known by the nickname 'Frankie Banana' due to his particular way of killing mobsters from rival gangs. Guess which way it might be? You guessed: bananas! Like the stockbroker, Frankie Banana was aware of that natural source of radioactivity: bananas. And in contrast to the cruel methods of his colleagues of Murder Inc., he preferred murder by bananas. It was a slow death, however, since the lethal dose of radiation poisoning is 10.000.000 bananas. In one fell swoop.

WHICH ATOM HAS ZERO NEUTRONS?

An atom without neutrons! What a load of garbage...! And yet that aberration exists. It occurs with hydrogen. Not with all the hydrogen, just some bastard atoms. If my advice had been followed, that would not have happened! But that arse-licker Harvard fool wanted to look good with Stanford and gave them a few hydrogen atoms. If we had given them colored marbles, as I suggested, they wouldn't have noticed the difference. "Hydrogen neutrons are very fast", we still warned them, "keep an eye on them." But they didn't heed the warning. The next day the provost summoned me to his office: "Hey, Sam, we got a call from Stanford. They say that the hydrogen atoms we gave them lack neutrons." "Damn it, they have already broken out! I did warn you!" "Are you sure those atoms had neutrons?" "What do you think I experiment with?! With chicken eggs?" Bureaucrats! I can't stand them!

WHICH EXPERIMENTAL DEMONSTRATION OF QUANTUM PRINCIPLES WAS DESCRIBED BY 'PHYSICS WORLD' AS "THE MOST BEAUTIFUL EXPERIMENT" OF ALL TIME?

In the course of my career as a particle physicist, I have performed lots of beautiful experiments. Surely, Physics World Magazine were referring to one of them. When I still was at Columbia, for example, I trained four electrons to choreograph the Dance of the Little Swans from Tchaikovsky's "Swan Lake". What took me the most work was that they learned to walk on their toes. But they did it, and received a rave from Physics World. In view of this, I tried to make them debut in a Broadway theater, but each spectator had to bring their own particle microscope and producers were not willing to take the risk. On another occasion, together with a group of Caltech physicists, we performed an experiment with photons consisting of matching them with fireflies. It was about demonstrating how photons interact with matter. But you never know when a photon will decide to behave like a wave or like a particle, and this time the "Physics World" review was not so favorable. Also, the fireflies had trouble keeping up with the photons.

WHY IS A NEUTRON NEUTRAL?

A recent Brookhaven National Laboratory's experiment with neutrons concludes that between a pizza with pepperoni and another pizza without pepperoni the neutron will always choose pizza with pepperoni. This has incalculable implications for particle physics. The neutron should be neutral as far as types of pizza are concerned. However, he prefers the pepperoni. Professor Viktor Lazler, from Stanford University, has since tried to play down the problem by saying that he also prefers pizza with pepperoni and that does not entail incalculable implications. (However, his wife has hinted that he is in a romantic relationship with their television set, with which she has surprised him embraced on several occasions.) For his part, Otto Frankel, from Harvard, has been experimenting with mascarpone, and in his Science Magazine's article "Pizza preference in Quantum Chromodynamics", states that the neutron does not reject pizza with mascarpone, but prefers the pepperoni even if the piece of pizza is smaller. Undoubtedly this means something, but it is still too soon to say what.

HOW WOULD YOU REACT IF YOU WITNESSED A PARANORMAL PHENOMENA THAT CONTRAVENED THE LAWS OF SCIENCE?

I would face this phenomenon bravely! I do not want to show off as a hero, but more than once I admonished a paranormal phenomenon to disappear and to follow the principles of Science. I will only mention one case that almost cost me my life. I worked at the Brookhaven National Laboratory in NY (My specialty is particle physics). One afternoon I stayed working late. Towards midnight I left the laboratory to urinate. Apparently there was no one else on my floor. While I was walking down the long corridor plunged into my thoughts, I raised my head and stopped suddenly when I saw a ... (It's hard for me to write it because I do not believe or ever believe in it) a ghost. I know he was a ghost because he was semi transparent, and he lacked feet, which nevertheless did not stop him from moving back and forth, right and left and even up and down. My mind is too rational to admit what I was seeing, so I ignored it and resumed the march. Then the ghost began to move to all sides as if he danced. I kindly asked him to let me pass because I had the urgent need to go to the restroom. But when I saw that he did not pay any attention to me, I became infuriated and began shouting at him with all my might to get out of my way and to get out of the way of Science since he was a non-existent creature, probably the fruit of my feverish mind because of excessive work. That visibly bothered him: he began to grumble and to separate his head from his body again and again. I promised myself not to spend so many hours in the lab, and began to move towards him as if the hallway was clear. But when I reached him I received a strong push that made me fall on my back. Despite the pain and considering the ridiculous and impossible of the situation, I began to laugh hysterically. The phantom imitated me and we stayed like that, laughing the two of us for about five minutes. Until I got tired and I told him the sad truth for him: that he did not exist! As he started to laugh even louder, I jumped up, grabbed a fire extinguisher and threw it at his head, which was still floating on his body. But the object bounced and hit me so hard that I lost consciousness. In the morning, the concierge found me lying in the middle of the corridor and revived me with blows. I told him that I had tripped over the extinguisher and

did not discuss the incident with
anyone. Not even with myself because,
after all, what can not be can not be
and it is also impossible.

MASS IS CONSIDERED TO BE A FORM OF ENERGY. CAN ELECTRICAL CHARGE ALSO BE CONSIDERED A FORM OF ENERGY?

Mass a form of Energy! Evidently, you have also been duped by that hilarious equation $E=mc^2$! Haven't you noticed Einstein was winking and laughing when he made public his famous theory? He made it public on stage at a popular San Francisco nightclub called "hungry i". He was performing one of his 'scientific' monologues. For some baffling reason most people take his jokes seriously. Everyone overlooks the fact that, first and foremost, he was a stand up comedian specialized in scientific humor. He had such good jokes! Like the one about the hydrogen atom which walks into a bar and says 'I've lost my electron'. 'Are you sure?'

asks the bartender. 'Yes, I'm positive'. Ha, ha, ha! What a hilarious chap he was!

WHEN THE LARGE HADRON COLLIDER WAS CREATED, THERE WAS AN EXPECTATION THAT IT WOULD LEAD TO THE DISCOVERY OF THE HIGGS BOSON. DO WE HAVE ANY SIMILAR DISCOVERIES WE ARE EXPECTING TO MAKE IN THE NEAR FUTURE?

Yes, indeed. The entire scientific community is on pins and needles waiting for me to discover a new particle called 'Perez-Pons tuson'. And while I'm at it, I plan to build an irrefutable Theory of Everything. But don't rush me, I like to be certain to cross every "t". However, I give you a preview: This quantum entity that I am about to discover not only may be described as either a particle or a wave, but also as a field-mouse running in a zigzag pattern. This is a scoop for you, I know, but try to keep it a secret until I find out how the hell this can be.

The thing is that right now I can't concentrate on my work. You know, lately I spend the nights without sleeping. I lie awake fearing that they will return at any time! ... Ah, but naturally you don't know what I'm talking about ... Let me tell you. But keep secret about this too:

Nine days ago I was trying to fall asleep after a hard day solving physics equations, when all of a sudden, plop, I was illuminated by a very powerful beam of light coming from above! For a moment I was dumbfounded thinking that they had just awarded me the Nobel Prize in Physics. I even got up and bowed ... until I started to rise upwards like a helium balloon! At that moment I realized that I was being abducted by aliens!

Then I found myself sitting in a barber chair surrounded by four macrocephalic little aliens with black bulging eyes. I heard the barber's scissors click in their thin hands and I shuddered. A stupid inertia prompted me to order the type of haircut I wanted: a serious and discreet cut as befits a scientist like me. That order of mine (emptied, with a part in the middle...) unleashed a huge hilarity among the aliens, which caused me a deep indignation. Scientists in general do not appreciate the sense of humor. The Universe does not joke!, as I usually say to my students. In a burst of self-love I got up and shouted that they were laughing at a reputable scientist who was looking for nothing less than the Theory of Everything. Then there was silence and immediately the laughter doubled. The aliens wallowed on the floor laughing with my growing indignation, when suddenly one of them wrote with his finger in the air a simple equation.

There it was: the long-awaited formula of the Grand Unified Theory!

Absorbed in its simplicity, I let the aliens escort me back to the barber chair and do with my hair the worst stylistic atrocities. They showed me a mirror so that I could see the results of their work, and they burst out laughing again at the expense of my hair. But I heard them from afar, I was focused on the formula. I remember that they accompanied me to the exit, where for the first time they became serious and then I heard inside my head these words: "We regret having to erase from your memory the Theory of Everything". No! I protested. But what was I going to do? It was four against one, and those four had ray guns! I was helpless, and I waited resigned to being injected with a serum for oblivion or something similar. But they just shook me a slap. When I woke up, I was in the middle of a field of wheat, where I realized with despair that the precious formula had disappeared without leaving a trace from my head of an iridescent blue.

ARE NEUTRONS ACTUALLY SPHERICAL?

Indeed, neutrons are spherical. But neutron's limbs are factors to be reckoned with. Neutrons' sphericity causes them to walk swaying and, when running, sometimes they fall, with the consequent danger to the scientists who work with them. In my MIT laboratory I have suffered the consequences of three neutron's falls. The first time I got fired out the window. The second time the neutron suffered a concussion. And the third time the entire laboratory was fired through the window and we both suffered a concussion. To get around those dangers it's necessary to prevent the neutron from running, which is what particle physics research is focused on now.

WHY IS AN ISOLATED NEUTRON UNSTABLE?

To answer this question intelligibly, we must clarify that there are two types of neutrons, fast and slow. As its name suggests, fast neutrons are capable of running at breakneck speed (especially when chased by an electron.) This is a serious inconvenience for scientists, since in conditions of freedom, there is a risk that suddenly the neutron starts running, being extremely difficult to catch it. That's why particle physicists prefer to work with slow neutrons. In 2012, Henri Magritte, from CERN, experimented with the possibility of shackling fast neutrons, thus turning them into slow neutrons. However, the experiment was a failure due to neutrons' ability (more suited to Harry Houdini) to get rid of the shackles.

WHAT NATURALLY OCURRING RADIOISOTOPE IS USED AS THE FUEL FOR A NUCLEAR REACTOR?

Bananas

WHAT IS BETWEEN ATOMS? IS IT JUST AN EMPTY SPACE?

When I gave University Lectures on the West Coast (now I am banned there because of an outlaw problem), I used to end my speech by proposing to the audience this same question: "What is between atoms? Is it just an empty space?". In the Pacific Coast people have the bizarre impression that 'full' and 'empty' are contradictory concepts, you know? "In particle physics (I explained) there is the notion of 'full of emptiness'. A graphic example will make that clear. Gentlemen, empty your pockets, would you?" I said, and then I took out a bag and urged them to put the content of their pockets in it without forgetting wallets. "Now check your pockets and you will understand the notion." And with this unexpected ending I concluded my conference and fled. That's how I made my fortune.

SENSATIONAL DISCOVERY!

A group of scientists working at the Research Institute of the Minnesota Lunatic Asylum have made a discovery that breaks the Physics' current theoretical paradigm. According to the theory prevailing to this day, the quark is the last piece of matter. There is nothing inside the quark except a little bit of suspended dust. Well, now this standard model has turned out to be out of date. Indeed, from the continuous cross-eyed observation of hadrons (composite particles), Samantha Philipovna, Paul Wagener and Tod 'Knucklehead' Benson have found, under the influence of alcohol, that inside a quark there is something. Or more precisely, someone. Yes, yes, you heard right: there is someone. To be more precise, a bald guy with glasses and a bushy beard. Of course, this is an extraordinary claim. And, as Carl Sagan put it, "extraordinary claims require extraordinary evidence". That's why at this very moment, the main research institutes around the world are trying to validate the discovery by reproducing the conditions in which it was made. Meanwhile, the whole scientific community (among which I include myself) is on pins and needles waiting for news.

THE MAN IN THE VERY CORE OF MATTER

"How the hell did that asshole get in there?!" was Dr. F. Gianotti's enraged reaction to the recent discovery of a bald man inside the most intimate center of the atom. This question asked in the last issue of the Science Magazine has upset some of Dr. Gianotti's colleagues, who frowned at her friend's outburst alleging that it's not edifying to disqualify the experimental data of Science. "It is not scientifically proven that Mr. Outspensky is an asshole" noted Spencer Ward from the University of Arizona. Yet while it is true that the physicists who have reproduced the conditions of the experiment (state of drunkenness and mental imbalance) unanimously agree that Mr.

Outspensky can give the impression of imbecility due to his exasperated way of gesturing, it is still too soon to assert this designation. In this regard, the Massachusetts Institute of Technology has allocated (at my request) thirty thousand dollars to an exhaustive investigation into the life of Gordon Outspensky, the man in the very core of Matter.

FIRST FINDINGS ABOUT G. OUTSPENSKY, THE-MAN-IN-THE-CORE-OF-MATTER

Gordon Outspensky was born in 1965 in Novosibirsk to a wealthy family of American emigrants from New Jersey. His father, Wolfgang, had emigrated to Siberia in search of bankruptcy because he didn't like bookkeeping. When he finally got ruined, the family returned to New Jersey, where the father set up a junk shop to get by, with which he made a fortune again. Disgusted, he returned to Siberia with his extended family. However, Gordon, who was already fifteen years old, wanted to stay in America because of his fondness for radioactivity, which he used to snort at a radioactive waste's landfill in his neighborhood. Researchers suspect that it was around this time that he began to get acquainted with the nuclear decay (the process by which an unstable atomic nucleus loses energy by radiation).

THE OUTSPENSKY ENIGMA

The commotion caused by the recent discovery of a New Jersey inhabitant inside what until now was considered the last piece of matter, continues. "Is this G. Outspensky the 'daughter nuclide' referred to in quantum theory?" is the question that Physics students from all over the world wonder. But their teachers reply that it is too hairy an individual to be a female. Furthermore, the Outspensky family has identified their relative definitely. Also, Mr. Outspensky's disappearance coincides in time with the discovery of "this bearded and gesticulating individual" within the most intimate interiority of the atom. The question is: how did he get there? Professor Jean Boulard from the CNRS has suggested an inconceivable possibility: "Could there be a wormhole that links the macrocosm with the microcosm? And if so, would travelers find any problem in acquiring supplies throughout the whole journey?" Naturally, such a suggestion has heated the atmosphere among the scientific community, which has begun to bear arms. (Last week a shooting starring various prestigious particle physicists was reported at Stanford University). Some scientists point to molecular nanotechnology to explain the Outspensky phenomenon. P. Fujikawa, from the Hitachi Central Research Laboratory, has been experimenting on himself with the aim of reducing his size to less than 100 nanometers. In one of his first attempts, he was caught trying to squeeze himself in a shoe box.

STARTLING REVELATIONS ABOUT THE-MAN-AT-THE-CORE-OF-MATTER

"He used to cast strange spells out of a book" is the last statement of Mrs. Millicent Barnaby, who was a neighbor of G. Outspensky in New Jersey for over twenty years. Reliable sources from the FBI assure that his old house has been thoroughly searched for clues that shed some light on this case that has the scientific community dangling. Officers were forced to wear radiation protective suits and helmets since detection instruments revealed high radioactivity levels throughout the house. They are particularly interested in Mr. Outspensky's library, almost entirely made up of books on quantum physics -a heavy reading matter for a daily labourer with a level of education no higher than that of kindergarten. After the neighbor's aforementioned statement, the interest is now focused on finding out which of those multitude of books is the mysterious one that allegedly G. Outspensky used for his spells. On the other hand, he has some criminal history but all related to minor crimes such as bypassing the ban on feeding zoo animals, or outlandish behavior. (He wore a pineapple embedded in his head as a hat, and did not allow anyone to order the same dish when he went to a restaurant.)

THE QUANTUM SPELLBOOK

After the exhaustive and unsuccessful search of Outspensky's library in quest of his mysterious spellbook, the FBI agents noticed that there was a small pile of ashes in the fireplace. They quickly deduced that before embarking on his journey into the innermost recesses of Matter, Outspensky had proceeded to dispose of the book by burning it. The ashes were immediately sent to an FBI laboratory, where an expert in document restoration pieced the powdery residues together, thus being able to completely recover the original book. This one turned out to be titled "Quantum Tunneling: now I'm here, now I'm there (Teleportation Manual)" and several copies have already been distributed for study among the most eminent specialists in Quantum Mechanics, including myself.

INCREDIBLE!

From the reading of Outspensky's mysterious burned book, it is inferred that all quantum physics is based on wrong vocabulary, which distorts all our knowledge about the ins and outs of the Universe. First of all, the real name of what particle physicists call 'quark' is 'pojk' (pronounced 'puagh'). The names of the six types of quarks (up, down, top, bottom, strange and charm) not only are they bizarre but totally unfounded: they must be called Elif, Mortimer, Aylin, Yachica, Maadai, and Yosef Tzvi Finkel. Instead of "Alpha particle" we should speak of "Dodyt particle", and gluons' right name is brriws (pronounced 'bubs'). The true name of gluino is 'brriyeah'. On the other hand, the lepton is actually the hadron and the hadron is actually the lepton. Boson is fermion and fermion is 'the one with the wart on her'. Moreover, the term SUSY to refer to supersymmetry not only is it wrong but also offensive: instead of 'SUSY' particle we should say 'Miss Susan Gladstone, of Philadelphia' particle. The names of meson (real name: 'gorwy') and baryon (real name: 'Binyamin Amsalem') are also wrong according to the book in question... In short, the list of errors is endless, and that explains why Physics has not yet developed a plausible Theory of Everything.

SCIENCE'S URGENT GOAL: EVICT THE SQUATTER!

The recent discovery of a bearded man with glasses housed in the quark (the last particle of the Standard Model of Matter) has captured the popular imagination to a degree few politicians have equaled. To think that in the very core of each one of the infinite atoms of Matter resides a human being with a name (Gordon Outspensky), an address (Newark, New Jersey) and a social security number, has so stunned pseudoscience's fans that numerous sects have already sprung up far and wide. This Outspensky's weird and growing cult has alarmed physicists, who have resolved to evict the squatter from what most of them believe to be the smallest building block of Matter. However, the question arises: Evict him, yes! But in what way? "Law enforcement?" I'm afraid that Law enforcement agencies have no jurisdiction in the innermost core of Matter. "Maybe using surgical forceps?" But the forceps would have to be infinitesimal! And not just the forceps: the surgeon himself should be at least dimension sized from 1 to 100 nanometers! Do any of you know a surgeon like that? Well, in any case, if anyone else has any other stupid suggestion ...

HOW CAN A BEAM OF LIGHT TRAVEL AROUND THE EARTH ROUGHLY 7 TIMES IN A SECOND AND A NEUTRON STAR CAN SUPPOSEDLY ROTATE 1000 TIMES IN A SECOND? WOULDN'T THE ROTATIONAL SPEED OF THE NEUTRON STAR EXCEED THE SPEED OF LIGHT?

When a subatomic particle reaches the speed of light, it is too late to try to stop it. It is more prudent to stay waiting for events, behind a curtain if possible (without the feet protruding). Most likely, the particle in question eventually will crash on its own and problem solved. If it does not crash, the problem does nothing but grow in a spiral known as "growth of the problem" or "quantum mess". In this case, the particle must be bombed, for which the intervention of the army is required, which is always a nuisance. Therefore, it is advisable not to allow an atomic particle to reach the speed of light even if it means temporarily staying in the dark.

HOW FAST ARE GLUONS EXCHANGED BETWEEN QUARKS? IS IT THE SPEED OF LIGHT OR SLOWER?

Sorry, but the question is wrongly formulated. The correct formulation would be: "Is Mickey Mouse an authentic mouse or is it a cartoon and we have been fooled?" If we have been fooled, the speed of the gluons exchanged between Mickey Mouse's quarks equals zero. The reason is that a cartoon lacks mass, therefore atoms, therefore neutrons, therefore quarks. A cartoon lacks everything! So it's a fake and you've the right to file a complaint with a court of law. But make sure it's an authentic court of law and not a cartoon, otherwise we'll have been fooled twice! (Paradoxically, the speed at which we are fooled can approach the speed of light.) Frankly, I don't know what I mean by all this.

IS THERE ANY GREAT WOMAN PHYSICIST LIKE EINSTEIN AND NEWTON?

Female physicists are not only as smart as their male peers but they have the added bonus that do not usually lost the keys nor wear their clothes inside out. (Einstein often had trouble going to the bathroom because he couldn't find his fly: he would start yelling that someone had sabotaged his pants until he was told that he wore the pants on backwards.) It would be easy to praise the merits of the two women who were awarded with the Nobel Prize in Physics. But I prefer to mention one of the many female particle physicist who would have deserved such distinction. Her name is Lucinda Shroedinger, whose mathematical model explained why certain number of nucleons do not collaborate in the maintenance of the atom and instead prefer to get drunk.

Thus shedding some light on the reason for the erratic behavior that characterizes the quantum world.

WHY DO THE ELECTRONS REVOLVE AROUND THE NUCLEUS?

Here's a smart question. Why?! Why?! Why does the electron revolve around the nucleus? Doesn't it have anything better to do? This is what I've been wondering since I am able to reason, that is, two years after being hired by MIT to take over the Laboratory for Nuclear Science. The only answer I can think of is the answer which is implicit in the very question: "Doesn't the electron have anything better to do than to revolve around the nucleus?" Well, maybe not. We've got to keep in mind that everything indicates that, within the atom, there is not much to entertain yourself with. As in many small towns, also in the atom all the fun is being had in the center or the nucleus. However, the atomic nucleus is so tiny that there is hardly room for a cinema or a dance hall, let alone a playing field or court of any kind. In such circumstances, revolving around the nucleus may be the best option to hang out.

WHEN AN ELECTRON IN A COMPOUND (SAY, A CHLOROPHYLL MOLECULE) IS EXCITED, WHICH ELECTRON ON WHICH ATOM IS MOVED TO A HIGHER ENERGY LEVEL?

When you say 'excited', do you mean 'sexually aroused'? In this case, the experiments carried out in our laboratory at MIT show that It is not an easy task to achieve a sufficient level of excitation in the electron for it to move to a higher energy level. Particularly in the case of chlorophyll molecules, electrons tend to behave with extreme shyness. You know, blushing, stammering and all that. Often they hide their heads under a layer of antimatter that protects them from unwanted fingering and other invasions of their privacy, such as pointless talk that leads to nothing. On such occasions the electron stops orbiting around the nucleus with the consequent danger of disintegration for the molecule. Therefore, particle physicists strongly advise against any kind of sexual provocation to which an electron may be subjected. Instead, they recommend platonic or chaste approaches, to which electrons are much more responsive.

WHO IS WORKING ON A THEORY OF EVERYTHING?

I am! However, to fit all physical phenomena so that they adapt to a Theory of Everything is not as simple as I had supposed. I have come to the conclusion that certain uncomfortable elements must be eliminated from the equation. For example, the Law of Gravity (a law very overvalued in my opinion). So that my colleagues do not accuse me of arbitrariness, I have not wanted to ignore the scientific method, and I have already taken twelve jumps from the second floor of my office. When I recover from my injuries and my bones finish welding I will keep trying. It would suffice only ONCE that, instead of falling, I would take flight, so that the damn Law of Gravity would be compromised. Thus, Science would be closer to discovering the desired Grand Unifying Theory. (If only I could sleep at night! But now it is not only the aliens who disturb me, but other weird creatures: the succubi! I wonder where these damned creatures fit into the Theory of Everything.)

DO WE ACTUALLY KNOW WHAT KEEPS ELECTRONS IN ORBIT AROUND THE NUCLEUS OF ATOMS? OR DO WE ONLY HAVE MODELS AND THEORIES TO HELP US IMAGINE WHAT'S GOING ON IN THERE?

Wow! Such a good question: what's going on in there? You should be awarded the Nobel Prize in Physics just for it. Anyway, the prize should be shared since I've been asking myself the same question. Only, I repeat it several times, also adding many exclamation points. If there was a punctuation mark in English grammar to indicate the state of despair depicted by Edvard Munch's painting "The Scream", I'd add that too. What the hell is going on in there?!!!!!!!!! What the hell is going on in there?!!!!!!!!! What the hell is going on in there?!!!!!!!!! What the hell is going on in there?!!!!!!!!! (In the event of winning the Nobel Prize, I want it on record that the addition of "hell" is also mine.)

WHY ARE PROTONS AND NEUTRONS MADE UP OF 3 QUARKS? WHY IS 3 THE MAGIC NUMBER?

To glimpse a quark you have to close your eyes. With open eyes they are impossible to see, unless you flip back looking out of the corner of your eye. But I broke some ribs when I tried it, so I strongly recommend closing eyes to see quarks. As for number 3, it depends on how you count. If you start counting from number 2, you get 4, which subtracted from three results in 1, which is the primary quark or antiquark. I've been conducting blind experiments with quarks in a bowling alley (I prefer it to particle accelerators) and I can say that the quark knocked down three pins, which is significant. Uh?

WHO ELIMINATED THE LIMITATIONS OF BOHR'S ATOMIC MODEL?

It was me. After two months searching around for various solutions, I had a spark moment and came up with an idea. To remove the limitations of the Bohr model, all that was needed was to grab the Bohr model and throw it in the trash. Thereafter I developed a new model without limitations: the 'somersault model' of the atom, according to which electrons go around in pairs doing somersaults, juggling five positrons at once and performing other slapstick actions, such as pratfalls. All this under the watchful gaze of the protons and neutrons of the nucleus. The astonishment caused within the nucleus by this show is what generates the different isotopes of a chemical element.

IN A HYDROGEN ATOM, ENERGY OF THE FIRST EXCITED STATE IS -3.4 ELECTRON VOLT, THEN WHAT IS THE KINETIC ENERGY OF THE SAME ORBIT OF A HYDROGEN ATOM?

That's a good question. The answer is "The battle of Stalingrad". I am aware that at first sight the question seems better than the answer. But that is because the question and the answer do not fit together. In itself, considered independently of your question, my answer is better than your question because it is shorter, more understandable and does not beat around the bush. And, above all, does not causes headache. So I'm proposing you to acknowledge defeat and to try to adapt your question to my answer. If you would like a suggestion, how about something like this: "What battle was fought in the city of Stalingrad?"?

A BUCKET IS FULL OF WATER. 50% OF THE WATER IS TAKEN FROM IT EVERY HOUR. HOW MANY HOURS WILL IT TAKE UNTIL IT IS COMPLETELY EMPTY?

This is the typical trick question. In particle physics there are a lot of trick questions and tricksters, and you have to walk on eggshells to avoid making a fool of yourself. To this end, you have to be faster than the trickster and get ahead of his question by reversing it. When you hear him say "A bucket is full of water…" you have to smell the trick question behind this preamble and talk back: "A water is full of buckets!" He'll ask "What water?" and you say "A molecule of water!". At this point the trickster will become cross-eyed by this unexpected turn around of the question. That is when you should take the opportunity to slap him, and drop that that's the correct answer to his question.

IN THE CONTEXT OF PET SCAN, FLUORINE-18 HAS 9 PROTONS AND 9 NEUTRONS NICELY BALANCED. WHY WOULD IT WANT TO UNDERGO BETA DECAY AND LOSE A PROTON, GAIN A NEUTRON TO BECOME OXYGEN-18 (8 PROTONS, 10 NEUTRONS WHICH SEEMS UNBALANCED)?

The answer to your question is "Daisy Petal Pickin' by Jimmy Gilmer & The Fireballs". Anyway, let me tell you that you shouldn't go around asking this kind of indiscreet question. If such a question falls into the wrong hands, it could be a danger to national security. There are certain issues that it is better not to bring up. The bizarre transformation of fluorine-18 into oxygen-18 is one of those issues. Boldly, you ask why. And I ask you in the name of Science: What difference does it make to you?! It's not you that's guilty! So why should it bother you?! If fluorine-18 take to becoming oxygen-18, it must have its reasons. And if not, worse for it. It's old enough to know what it's doing. My cousin Raphael took to becoming 'busty Lolita'. One day at a family celebration I asked my aunt who was that woman with such big breasts. "Don't you recognize your cousin Raphael?" she exclaimed. Since then I have learned not to ask indiscreet questions.

WHY DO ACCELERATED ELECTRONS RADIATE ENERGY, I. E., IN PARTICLE ACCELERATORS, AND DO THEY LOSE THEIR EFFECTIVE MASS AS A RESULT?

I am glad that this question is on today's agenda. I've been accelerating electrons in my MIT lab for two years now (I don't like the particle accelerator because it spoils particles). I will explain how I do it in case you want to do it yourself at home. To shoot the electron correctly, you have to round your index finger around the electron and put your thumb behind it, and then push the thumb forward. You will see the electron shoot out like a torpedo. Well, that being said, I answer the question. As my mentor Albert Einstein wisely said: "If you can't explain something in simple terms, that means that you don't understand it enough." So I will be very clear and concise. Accelerated electrons radiate energy because they lose their mass and they lose their mass because they radiate energy. Plain and simple.

WHAT IS A QUARK FIELD?

A quark field is the field in which each type (or flavor) of quark plays ball, whether football or basketball. Since quarks generally have a spherical shape, accidents abound. Indeed, according to the 'quantum field theory', when a quark is mistaken for the ball, the kicked or thrown quark becomes enraged, generating a chain reaction in which all quarks in the field begin to kick each other and be thrown into the basket, what is known as "quantum mess". Under these circumstances, the atom overheats, and may explode. However, the most common consequence is that it just begins to behave erratically (for example: singing Tyrolean melodies or/and pretending to be a duck).

WHAT IS A LAYMAN'S EXPLANATION OF THE STANDARD MODEL OF PARTICLE PHYSICS?

Uh! the Standard Model! It's got so many gaps that you can't trust it unless you have a good lawyer. And even then, most likely, not only you are going to be scammed, but also your lawyer, and your lawyer's in-laws, and the cousins of your lawyer's in-laws are going to be scammed. A lot of people, starting with the particle physicists, have been scammed by the Standard Model since its birth in Mollerussa to a family of paper-dress tailors. It leaves so many phenomena unexplained! Dark matter, gravitation, baryon asimmetry, neutrino oscillations, the accelerating expansion of the Universe, and where does uncle Fischel go every Thursday night when aunt Rifka falls asleep (he leaves home naked and returns shortly before dawn and then puts on his pajamas and goes back to bed). None of these mysteries can be explained by the Standard Model. That's why I am trying to introduce other variables into the equation, such as the surprise factor.

WHAT IS A BLACK HOLE?

This whole black hole thing blows me away! It is time for this unfortunate misunderstanding to be undone. Pay heed. After the theoretical discovery of black holes, scientists from around the planet set out to look for a black hole. But that a thing exists theoretically does not imply that it must also exist in real life. Theoretically I am one of the smartest scientists in the world. In practice, however, I am stupid, as you can see for yourself. So much so that I thought I had found a black hole myself. Concretely on a farm in Pennsylvania. I had never been to a farm. There I saw incredible things. You'd never believe it. Do you know where they get the milk you buy in the supermarket. Not to mention the eggs! But back to the case here... There was a black hole in the middle of the farm yard. A black hole dropped from Space! -that's what I thought. Its tremendous gravitational force, coupled with my insatiable scientific curiosity, irresistibly attracted me to it. I rushed to lean over the circular low wall that surrounded it and immediately the hole swallowed my hat and glasses. I focused its blackness with my flashlight and it swallowed my flashlight too. "Aha!" I thought, "the theory is right: black holes swallow the light!" But then the farmer came running and scold me for throwing things into the well. Man!, you'd be amazed at where they get water from in rural areas.

HOW ARE NEUTRONS AND PROTONS DIFFERENT?

Both are nucleons. But you can not rely on protons. First thing you know, they will bind to electrons. Even lacking electric charge, a neutron never will leave you in the lurch. It will not run after no electron even though it is a free neutron. Maybe it would like to be engaged to its sweetheart, but neutron is chaste. Proton is promiscuous. That's another difference. Certainly, at some point the neutron languishes out of nostalgia for its sweetheart and decays transforming itself into a proton. However, before this, the neutron performs a solo dance called "The Dying Swan". This dance is very touching: the neutron waves its arms while taking tiny steps called "pas de bourrée suivi". The French particle physicist André Levinson describes it this way: "Arms folded, on tiptoe, the neutron dreamily and slowly circles the atom. By even, gliding motions of the hands, returning to the background from whence it emerged, the neutron seems to strive toward the horizon, as though a moment more and it will fly—exploring the confines of space with its soul. The tension gradually relaxes and neutron sinks to earth, arms waving faintly as in pain. Then faltering with irregular steps toward the edge of the atom—leg bones quiver like the strings of a harp—by one swift forward-gliding motion of the right foot to earth, it sinks on the left knee—the aerial creature struggling against earthly bonds; and there, transfixed by pain, the neutron dies."

CAN THERE BE AN -ONIUM OF A NEUTRON AND AN ANTI-NEUTRON?

This is a stupid question. Not only the question is stupid but the answer to the question is stupid and the whole context surrounding the question is stupid. Anyway I'll try to put some clarity on so much stupidity. Let's see. To get started, the assumption you raise requires a 'bound state' (or marriage) between a particle and its antiparticle, which, although possible in reality, is theoretically impossible. This impossible possibility is called 'onium' (not onion, as those stupid students insist on saying). In the case of neutron and anti-neutron, we would be talking about a neutronium, which is something completely different, so the example is a stupidity also. So let's take another example: the elementary particle called 'muon'. If we take a muon and an anti-muon and force them to get married, we will get a muonium and a big headache since a muonium is not a muon-antimuon marriage but an electron-antimuon marriage. (This didn't happen when it was the parents who chose their offsprings' spouse, and not the offsprings who married who they wanted.) The muon-antimuon marriage is called 'true muonium' or 'matrimonium', and that doesnt admit divorce or any kind of shady deal.

CAN YOU EXPLAIN IN SIMPLE TERMS THE QUANTUM-MECHANICAL IMPLEMENTATION OF THE 'SHELL GAME'?

No, I can't! In any case, it has to do with how light and matter interact. Because it is not only that light illuminates matter, you know? If only it were that simple! Where a layman sees light and sees matter, a scientist sees photons and sees electrons, neutrons and protons. This is most unfortunate but it's the case. I remember when I was a sophomore at Columbia University. I was dating a girl named Jennifer Solinsky. My friends told me that she had beautiful blue eyes, but I only saw a lot of mass full of particles. "You have some nice protons," I told her one day. Then I felt a strong itch on my cheek and lost consciousness. The next thing I remember is waking up in a bed at

Mount Sinai Hospital. I've never heard from Jennifer Solinsky since then.

CAN NON-EXISTENCE REALLY EXIST?

As a scientist I have been asking this question for a while and there is only one possible answer: what does not exist does not exist, period. Do not be fooled by apparently existing phenomena. The non-existent PRETEND to exist but in reality it does not exist no matter how hard it tries. If you are in one of these anomalous situations and have any hesitation, consult the catalog of what exists according to Science. If that strange phenomenon that you are witnessing does not appear in the catalog, IT DOES NOT EXIST. IT PRETENDS TO EXIST. The non-existent may not exist but it is not stupid, it is cunning and will try by all means to make you believe that it exists. But do not be fooled: pretend you too, bluff, look the other way, pretend not to notice. It's what I do constantly. Look, when I resided in New York, I was living in the same Queens' apartment for 15 years, and I was constantly the victim of one of these stubborn non-existent phenomena. The objects were blown up, the furniture crawled on the floor, the dishes were thrown out of the closet. Do you think this did not bother me? Of course! But I bore it in the name of Science! The laws of Science are to be obeyed. If those laws are broken, what do we get? A chaos! A chaos like the one that formed every time I was at home. Because of this damn chaos, I was on sick leave. Not that any of the objects that continually fluttered around me had impacted me (although countless times the coffee maker or the toaster passed less than an inch from my head). No. One day I met in the elevator with my downstairs neighbor, a woman strong as a bull. She said I dragged furniture up and down my apartment at night and didn't let her sleep. My explanation of non-existence did not seem to convince her to judge by the beating she gave me. But do you think these setbacks were going to make me move out of my apartment? Never! That would've been tantamount to tacitly admitting the existence of the non-existent, which would've meant a humiliation for Science. Let the dishes fly around you as much as they want! Do you think a scientist is going to be impressed by something that can not be happening ?! Do you underestimate Science so much?! It's just a matter of psyching yourself up ... and avoiding crossing paths with the neighbor below.

WHEN IS IT THOUGHT THAT THE 'THEORY OF EVERYTHING' WILL BE DISCOVERED AND WELL-DEFINED?

The "Everything" of the Theory of Everything (TOE) encompasses too many stuff to be able to integrate them all into one theory. Not to mention the hair dyeing aliens and the lustful succubi, just to name two weird phenomena of which I am a victim myself. But then there are those tiny mushroom-shaped creatures that chase Brad Connolly around the MIT campus in order to pickle him in brine. The fact that only he can see them does not mean that they are only in his mind. After all, his IQ is 160. His mind must be able to glimpse things that others could not even dream of. And then there is Professor Minkowski who says that his computer sends him love messages.

How the hell does that fit into the Theory of Everything? I wonder what Minkowski's IQ is.

HOW DOES AN ANTI-NEUTRON WORK?

A neutron is an uncharged particle, which means that neutron is an uncharged particle, which means exactly that. This may seem silly and that is exactly what it is, but now imagine that someone says: "A neutron is a charged particle". And he says this not because his life depends on it, but just because he is determined to be contrary, even if that gets him strung up from a tree. Well, that's an anti-neutron: a pain in the ass! Excuse my language but I can't stand anti-neutrons, they just gargle my balls. Some of my MIT colleagues (let them burn in hell) are aware of this and tend to provoke me by throwing anti-neutrons into my lab through the window. (My window lacks pane because, according to the provost - confound him!, the global budget of MIT would not be enough to replace my broken window pane every time there is an explosion in my laboratory. Damned bureaucrats!)

IN LAYPERSON'S TERMS, WHAT ARE THE CURRENT SPECIFIC OBSTACLES TO HAVING A GRAND UNIFIED THEORY OF PHYSICS? WHAT IS THE PROGRESS ON EACH OF THEM?

I'm working on it, don't rush me! I just took a new approach. It has to do with the shell game. I have the impression that this is going to be the only way to put quantum mechanics and special relativity together without them quarrelling. After all, it will turn out that Einstein was right: God does not play dice, He plays shell game. The great thing about this research is that it does not require highly sophisticated equipment. No particle accelerator or anything like that. Three walnut shells and a photon are enough. However, can you believe that the MIT provost has refused to finance the project? That man hates me. Although I must say that the feeling is mutual. Damn bureaucrats!

HOW DO ELECTRONS BEHAVE INSIDE THEIR SHAPES?

It was about time for someone to ask about this. Electrons seem to think anything is permissible when you are negatively charged. My cousin is negatively charged. However, he refrains from disproving theories of hard-working scientists like John Dalton! Do you know the time Mr. Dalton had invested in building up his atomic theory? He didn't even go on his honeymoon to Venice because he wanted to put to it the finishing touches. (His wife had to go without him, but she took revenge by sewing to the back the sleeves of all his shirts) And so much hard work for what? For nothing. For the electron to come and to tear down everything! The positron would never do that. Why don't electrons take the example from positrons? Here's a pertinent question.

HOW DOES QUANTUM FIELD THEORY ADDRESS WAVE-PARTICLE DUALITY?

With bewilderment. Quantum field theorists don't understand why light behaves in such a weird way. There is an element of unpredictability in light that keep them always on their toes, especially when they are having guests for dinner. They can never be sure how light will behave, whether properly, as particle, or as wave messing everything up. That's why quantum field theorists often prefer to dine in the dark, with the consequent danger of sticking the fork in any part of their faces. They live scared of light like the vampires, leaving their homes only at night and roaming the dimly lit streets of the suburbs in order to collect isotopes of different number of neutrons in each atom.

IS NON-EXISTENCE ALWAYS PREFERABLE TO EXISTENCE?

Now I'm not going to talk to you as a scientist, but as an existentialist philosopher: The non-existent exists, only that it exists in a non-existential or nonexistent way. So it is difficult to decide if non-existence is preferable to existence, since on the other hand existence can also be a nonexistent existence. In any case, if there were no other alternative, I would prefer existing existence.

HOW ARE NEUTRON STARS FORMED?

First of all, they have to learn music theory, tap dance and, if possible, to play an instrument. All that stuff is learned in the nucleus. When one neutron is hopeless at singing and dancing, can try to become a stand-up comedian. If it isn't funny either, it works in juggling pumpkins. In case of being a waste of space, it is dedicated to inducing nuclear fission. Generally, neutrons have a squeaky voice, that's why they often sing in playback or lip synch. This is frowned upon among particle physicists, but we condescend and do not deny them a chance. Now, the rise to stardom is never easy and there is a lot of competition between neutrons. Only a few of all neutrons that pursue stardom reach what, in quantum slang, is called their "neutron magnetic moment". Sometimes, all the neutrons of an atom associate and form an "atomic ensemble" in which everyone plays an instrument. The "neutron number" of an atomic ensemble may differ. The different variants (duo, trio, quartet, quintet ... and so on) are called "isotopes". Isotopes' favorite style is bebop.

WHAT IS THE MOST ACCURATE OF THE ATOMIC MODELS – THE BOHR MODEL OF CHEMISTRY, OR THE QUANTUM MECHANICAL MODEL OF PHYSICS?

I don't know whether it is the most accurate, but in any case the most outstanding model is the Perez-Pons model or myself model. According to this model that I have been outlining in recent years, the electron does not orbit around the nucleus (Bohr model) nor does it leap from one energy level to another (Quantum jumping model). The electron is not a kangaroo, gentlemen, let's be serious! What the electron actually does is play dead and then scare to death the dupe who dares to look at it. This I have demonstrated mathematically, and now I am eagerly awaiting for someone to experimentally validate it. If anyone hears of a researcher dropped dead while looking at an electron, please let me know. Don't make me wait long because I want to win the Nobel Prize in Physics and I have no time to lose.

WHY ARE ALIENS NEVER SEEN BY SCIENTISTS? IS THERE ANY PROOF OR EVIDENCE OF ALIEN/EXTRATERRESTRIAL LIFE EXISTENCE?

You offend me, ma'am! I am a prestigious scientist. And not only the aliens have allowed me to see them but they have dyed my hair a fluorescent blue color and they have left me bangs! And then they started laughing while they pointed me out, which for a reputable scientist is an intolerable humiliation. This has happened to me thirteen times so far. I urge the competent authorities to take measures to stop these attacks on good taste. (Is for this that I pay my taxes?!)

Excuse, ma'am, but your question makes me laugh. Ha, ha, ha, it's as if you ask a milkshake whether there are milkshake drinkers. Of course they exist! I have been abducted 13 times, like I said, without counting other frustrated attempts (once I cheated on them by pretending to be a refrigerator, ha, ha, ha, I'm very proud of that). I explain to you how it works: You are so quiet watching TV or sitting in the park. Suddenly, plop, a powerful beam of light illuminates you, as if you were on a stage under a spotlight. And you go up at great speed. It's them, the aliens, who from their damn flying saucer sip you like with a straw! They sip and sip until you fall inside the flying saucer, where, if you are bald, you have an implant of hair and, if you are not, they make you a bizarre hairstyle and they dye it blue, but a blue iridescent, an unknown blue here in the Earth. Then they laugh at your appearance.

HOW WAS THE HIGGS BOSON DISCOVERED?

Back in the '70s, three friends (François Englert, Robert Brout and myself) were committed to finding the "Higgs boson". For us it was the "Peter boson" because the three of us were Peter Higgs' close friends. Peter used to sit in the crying corner all day long. It was a very sad scene to witness. He just hid in a corner and wept because he had predicted the existence of a new subatomic particle called the Higgs boson but had no material evidence. So in solidarity with him, Robert, François and I set out to find that particle. The three of us used to meet in the evenings at François' house in Brussels. First we build a rudimentary particle collider. Then Robert developed a witty technique of particle launching: He held the particle up high against the top corner of his knuckle and then he kicked the thumb out moving it under the particle, thus causing this one to spin backwards as it was projected forwards. He taught us his technique and soon the three of us became expert particle launchers. Robert shot all types of subatomic particles. François preferred top quarks and I liked strange quarks better. We were colliding quarks at a higher and higher speed until one day the miracle happened: two quarks collided at such a speed that the collision produced a new particle. We'd just discovered the Higgs boson!

WHO REALLY UNDERSTANDS STRING THEORY?

I do! I am a string expert since once escaped from a federal prison down from a window by a string which I made myself with my own braided hair that I had been growing for years. While I braided my hair, inspiration struck me and I started to mentally formulate the String Theory. Such theory is based on the proven fact that there may be (although it's highly unlikely) an elementary particle called graviton. Another evidence that supports the soundness of the String Theory is the remote possibility that infinity has a real existence. Another irrefutable assertion of String Theory is that, under certain hypothetical conditions, it might be the case that there were multiple universes. It could even happen that these infinite universes were sublet and the rental contract wouldn't expire for at least an infinite number of years. As you see, String Theory rests on solid foundations and there is no doubt that it may be true. Or maybe not, who knows?

IF PARTICLES AND ANTIPARTICLES DESTROY EACH OTHER UPON INTERACTION, HOW DO A QUARK AND AN ANTIQUARK CREATE A MESON?

Do I have to remind you the "Bees and the Birds talk"? Save me the trouble, I beg you! I prefer to tell you a few words about particles and antiparticles if you don't mind. I hope you don't take this as an example of "You ask about whatever you want and I'll answer about whatever I please" (although that's exactly what it is).

If we had a microscope capable of showing us the inmost interiors of the Universe (That's to say a transmission electron microscope like the one we have in the MIT -but the provost withdrew my permission to use it because of an experiment that I was carrying out with a drunk electron. Damn provost!) If we had such microscope, I say, we would see with astonishment that the most basic ingredients of everything in the Universe (particles) have their 'anti'. A neutron has its antineutron, a boson, its antiboson; a fermion, its antifernion… Well, for some inextricable reason, each particle and its antiparticle mutually annihilate: both collide and disappear! It is as if the Universe took pleasure in revolting against itself. And I wonder: Is that a way to behave? (Every time I ask this question in the faculty, sparks fly and then I receive anonymous telephoned threats.)

Now, the important point is that at the initial moment of the Universe's emergence, there were as many particles as there were antiparticles. And they did what they do, that is, to collide and annihilate one another. Obviously, the foreseeable result should be that, moments after starting to exist, the Universe went fuck itself. Well, no! Good luck! It turns out that, somehow, the original balance between particles and antiparticles (between matter and antimatter) was tilted in favor of matter! At the end of this symmetric battle, there was still one last man standing! Can you believed it?! My colleagues are puzzled! They think that it is a magic trick and spend the whole day looking under the table, checking that the cards are not marked, looking for a false bottom in the hat ... But until now they have not figured out the trick. And yet, in the middle of the stage there is that huge white rabbit! Oh my God!

WHAT IS YOUR ALIEN STORY?

It's not just a story -it's a long history by now. I will tell you about the last episode, which happened the day before yesterday.

I was in my laboratory at the MIT dissecting a neutron, when suddenly a very powerful light burst through the window shinning a spotlight on me. This time I thought it was the TV show "Dancing with the Stars" and I started to dance wildly. But after a while I realized that I was floating, I looked down and saw that I had somehow left the MIT building and was being absorbed through a beam of light. Suddenly, I found myself surrounded by small, big-headed guys who were very thin and had big black eyes. They were green and had no teeth. I know this because they had their mouths wide open because of the laughter. I have to say that I kept dancing but it was because of my nerves. Then I understood that they laughed hysterically at my dance. I tried to stop, but I couldn't. I shimmied with my shoulders and shook like a bowl of jelly because I was scared to death, I must confess. After a temporary space that I would have no way to tell, I suddenly stopped when I received a heavy blow to the head. When I regained my senses, I was sitting in a barber chair and a few little green men were manipulating my hair while laughing. I have to say that aliens have a lot of sense of humor, that's for sure. The next thing I remember is being back in my MIT lab sitting on the floor. I had an urgent need to urinate and ran to the toilet. There I discovered in the mirror that my hair was dyed blue. An unknown blue here on Earth. In general, it was a very humiliating experience, especially for a scientist.

CAN YOU PLEASE TELL ME WHAT IS MAGIC NUMBERS WITH RESPECT TO ATOMIC NUCLEUS IN SIMPLE WORDS?

Uh, the magic number... It is three! I'm going to tell you how I discovered it. At that time I worked at the Brookhaven National Laboratory. I had just graduated from Columbia University where I earned a magna cum laude degree, plus a gigantic teddy bear in a raffle. I used to stay late at night researching with an Helium atom bought off at the black market. (At that time there was a shortage of Helium atoms due to the popularity of Donald Duck: in order to talk like the cartoon character, people inhaled helium, thus depleting stocks.) Well, I was hell-bent on making the atom jump through a hoop. That would be a great advance for Science: a trained atom! Next I would teach it to dance tap and sing Broadway melodies. That would make me win the Nobel Prize in Physics. (I was an ambitious scientist, and also had a gigantic teddy bear to support.) Allez hop!" I said, but the atom didn't flinch. "Allez hop!" I repeated. This time I saw the atom shake. "Allez hop!", I insisted, and this time succeeded: the atom jumped through the hoop! I had mastered matter! And then I laughed hysterically as mad scientists usually do.

WHAT NAME DID PARTICLE PHYSICIST RICHARD FEYNMAN USE WHILE REFERRING TO FEYNMAN DIAGRAMS?

'Myself diagrams'. Feynman was a tremendous particle collider. He was able to make collide three elementary particles at once with a single shot, knocking them out of the circle. He used to start the experiment with a single shooter proton and ended up with a shoebox filled to overflowing with quarks and gluons. He was undoubtedly a genius. Was also very good in mathematics: he kept a strict accounting not only of each one of the shots and their results, but of the trajectory and approximate speed that each particle had taken when pulled out of the circle. To this end, he used to draw complex color diagrams full of lines of different shapes -basically straight, dotted and squiggly. It is these diagrams that would end up being called 'Feynman diagrams'.

THE MESS AT THE HEART OF PHYSICS

This thing that looks like a bedside table might not be a bedside table. What's more, what is certain is that it is not a bedside table whatsoever. It could be anything except a bedside table. Such is a graphic description of the work in which I've been involved for twenty-two years! Do you know what that means in terms of emotional stress? Suppose your job is clerk in an office. You are typing on the computer. Then you look away for a moment, or just blink, and when you look at the screen again, there is no screen but you are milking a cow instead. Such is the level of uncertainty that particle physicists face daily! Now tell me whether it's strange or not that the other day I was arrested on the rooftop of the Green Building for shouting slogans in favor of the Bolivarian Revolution through a bullhorn.

WHAT IS THE MOST BEAUTIFUL EQUATION?

I will answer you through a short story included in my book "Silly Humor for Smart People"… Go ahead, buy it if you dare! (Coward!) The answer is right at the end of the story, which is entitled WAS ALBERT EINSTEIN A REPTILIAN?

In their office of ENIGMA CONSULTANTS S. L., Michael Schlimazl and Jacob Schlemihl deal with a client who asks them if they believe in extraterrestrials.

"Extraterrestrials? I'm afraid not. I'm not even sure the Earth is round. Half of the world's population would be living upside down. Do you know the headache that would cause that?"

"I think the Earth is shaped like a ball", Jacob answers in turn.

"You see?, your friend is better informed"

"He means an American football"

"Enough of nonsense. I don't care about the stupid things you believe or disbelieve in. The point is that I am offering you a case on a tray in the name of the president Roosevelt. Take it or leave it"

"Find out whether Albert Einstein is an extraterrestrial? And why has the president thought of us for that task? What about the CIA or the FBI?"

"Or the NFL?", Jacob adds.

"Jacob, Einstein is not the Giants' new quarterback. He is a reputable scientist." And turning to their client: "Why does the president suspect him?"

"You see, the president is very stressed lately. He discovers threats everywhere. There are the nazis, the communists,

Eleanor Roosevelt, that imbecile who occasionally walks in his underwear in front of the White House..." Suddenly he stares at Jacob, who blushes. "By the way, you look a lot like him, you wear the same hood" "Millions of people wear this kind of hood" replies Jacob. The client continues: "And now he has received a confidential report where they talk about a possible extraterrestrial infiltration on Earth. Specifically they talk of reptilian beings."

"What do you mean by reptilian?"

"They take the form of humans but they are a sort of lizards. They are recognized because, just like the reptiles, they have a long tongue that they constantly stick out."

"And why do you suspect Einstein?"

"When you see him, you'll understand. I made an appointment with him on your behalf. You have a scheduled lunch tomorrow at one o'clock at Delmonico's."

Jacob jumps: "At Delmonico's?! Who will pay the bill?!"

"We have already taken on that. Your task is to discover how truth is the president's suspicion. You'll pretend to be astronomers seeking life on other planets. That will serve as an alibi to ask the right questions."

The next day, Michael and Jacob show up elegantly dressed at "Delmonico's" reception.

"Table for two, gentlemen?"

"No, we wait for another person: Mr. Einstein"

"Mr. Einstein has already arrived." And turning to a waiter: "Ralph, accompany the gentlemen to table 5."

On table 5 sits Albert Einstein with his typical frizzy hair and protruding eyes, and sticking his tongue out so that it reaches up to his chin.

An hour later they have finished eating and they chat amiably.

"So you are looking for extraterrestrial life!"

"Correct", Jacob confirms with pride.

"Most likely, it does exist. But also most likely, you won't find it!"

"Why not?" Jacob protests. But Michael intervenes in support of Einstein:

"Mr. Einstein is right. You never find anything. You are always forced to wear a sock of another color"

"Oh, I don't question your ability", Einstein explains, "It's just that, in the case of existing, it's probably too far away, in another galaxy perhaps. Your telescopes don't reach"

"We don't use telescopes" Jacob points out.

"Astronomers without a telescope?", Einstein exclaims surprised.

"We have good eyesight."

Einstein bursts out laughing: "How funny are you!"

"Tell us, Mr. Einstein: What planet… I mean what city were you born?

Einstein doesn't stop laughing.

"Was it your mother who gave birth to you?"

"Ha, ha, ha! You can not be scientists. Scientists are so boring!"

Both stick out their tongues imitating their guest, who when is not speaking always has his tongue out reaching up to his chin.

"Ha, ha, ha! I see! You are clowns! Clowns are great imitators. I love clowns. But if you want to imitate me, let me make some tweaks.

He gets up and shakes their mops of hair until they are frizzy like his own. Then he sits down again: "Now we all look alike".

"Why is it that you always have your tongue out, if I may ask?"

"Oh, it's the force of habit. In Switzerland I worked for years in the postal service, pasting stamps."

Michael stands up and exclaims: "Well, that explains it, then. Mr. Einstein, it was nice meeting you, but we've got some business."

Jacob rises too and all shake hands.

"Sure, your show is in the afternoon, I guess. I would like to go see you. I love circus, but I am afraid that I have an appointment with some authentic scientists. Serious people, you know (He winks at them) Tedious people. Since I formulated that equation, they don't leave me alone!"

"What's that equation if it's not indiscreet?" asks Michael.

"$E = mc2$"

"Nice equation!"

"Yes, very beautiful" Jacob agrees.

"You should dedicate yourself to formulate equations and leave Science. That way you wouldn't have to deal with so many boring people."

Einstein bursts out laughing: "I'll think about it."

THE STRANGE QUARKS

One of the constant concerns of a particle physicist is the possibility of running into a quark of the strange or s type. One of my MIT colleagues, aware of my sympathy for neutrons, gave me a neutron that he had been working with lately. "Here you go, monkey (they call me 'monkey' because I walk on all fours due to a bet concerning Darwin's theory of evolution), I give it to you. I don't know how you can be so patient with these (and here he said a bad word that I won't reproduce)."

I locked myself in my lab and watched the neutron closely. (Actually, particles of atoms are not visible to the human eye, so I observe them with my eyes closed.) I immediately saw what the problem was: the neutron contained a strange or s quark.

First, it behaved like a boxer in the ring: it hopped around provoking the other quarks, until a top or t quark punched him and knocked him down. Then he began to make terrible faces imitating The Exorcist's girl, shouting obscenities to the other quarks and spewing them and vomiting them... Naturally, this destabilized the neutron, with the consequent risk to the atom, which, to make matters worse, had measles and whooping cough.

In cases like this, there is an action protocol consisting of putting the neutron in a cocktail's recipient and shaking it for five minutes. There is no quark that resists such a measure. I closed my eyes and looked again at the neutron and, in effect, the quark was dizzy.

Then one must proceed with the velocity of lightning. It is about inoculating a brigade of "police quarks" into the neutron so that they grab the s quark by the nuts and take it out of the neutron. (This must be made quickly because, in case the s quark recovers before leaving the neutron, it could be catastrophic since its mass is one hundred times larger than that of the "police quarks", which are usually of the up and down or u and d type.)

Once outside the neutron, the s quark is taken over by the researcher. (Here the danger is equal to zero because the mass of the researcher is a million times larger than that of the quark.)

Well, in the case at hand, I followed the protocol closely. But when I had the quark in my possession, it stared at me with a devilish look and let out a diabolical laugh that made my hair stand up. Then from its dirty mouth came a sequence of insults and obscenities, the slightest of which was "Your mother is a slut!" All said with such fury and with such a deep voice (so unusual in a sane quark, which voice is rather screeching) that my hair took off like a rocket, sticking itself into the lab's ceiling and leaving me bald-headed.

I was dumbstruck when suddenly the whole quark turned into a mouth and from that hole sprang a jet of vomit that hit me in full face. At this point, I couldn't help letting go the quark, which escaped.

This happened on February 4, 2016. Everything is recorded in the secret files of MIT, and my hair has never grown back.

CHARM QUARKS

I've been spying on a charm quark lately and I don't like what I have seen at all. Wouldn't surprise me if it got a neutrino pregnant. I just wrote an article about it for Science Magazine. What a dissolute life! If we allow such behavior to exist in the very heart of Matter, we are inevitably heading for disaster or, if not to disaster, at least to total extinction. Scientific authorities should do something about it. How is it possible that they have not noticed such promiscuous behavior on the part of this flavor of quark? I bet they've been looking the other way. ISC members! You'd better listen: we must root out this evil before it spreads to other elementary particles Otherwise, we'll find the atom turned into a brothel! I propose the implementation of drastic measures. Such as setting a curfew. A decent quark should be back in its atom before ten o'clock at night or suffer severe punishment, such as a drop in its allowance, or the ban of nights-out as long as it doesn't learn to behave like a gentle particle. We know, we know: by its very nature, Matter tends to worldly pleasures. But Spirit is there for something!

As you can see, I am truly indignant. And to top it off, succubi never stop bothering me. A scientist should be able to sleep at night without someone trying to rape him, I dare say!

WHAT IS INSIDE A QUARK?

Inside a quark there is a guy from New Jersey named Gordon Outspensky, but nobody knows how he got there. Scientists do not agree. Also, this Outspensky wears glasses, which contradicts everything that is known about quantum physics. Peter Higgs affirms that Outspensky's lenses cannot be made of organic glass because that would entail a cosmic cataclysm which would include the destruction of the Universe or, at least, of a part of New Jersey. In any case (according to Higgs) they must be made of antimatter glass. However, Ms. Outspensky assures that her husband commissioned his prescription lenses to a Newark optical shop called "Rainbow Glasses", where no one has the slightest idea what antimatter is, not even matter. In short, this is an unfathomable mystery that only the discovery of a Theory of Everything can solve.

NIGHT VISITORS

Tonight I was awakened by the noise of a racket in the living room. I was immediately alarmed at this because I had left some neutrons soaking overnight for an experiment that I planned to carry out in the morning. I quickly took my shotgun, jumped out of bed and rushed in the living room like a tornado. If you are a rational and sensible man like myself, you will not believe what my eyes saw. I didn't believe it either. And yet there they were: a succubus and an alien quarrelling over who had come first and therefore had the right to abduct me (the alien) or rape me (the succubus). Can there be greater humiliation for a scientist? I immediately pulled my shotgun on them. However, they continued fighting just like that. Then the anger took over me and I started firing my weapon aimlessly with a spray of bullets in the air (due to my squinting I have some difficulty aiming at a specific target). The next thing I remember is waking up at dawn. There I was, lying on the floor with the shotgun next to me. I was naked, and the neutrons I had left soaking overnight were conspicuous by being absent. Terrified at the prospect that the delicate particles had fallen into the wrong hands, I went straight to the provost without even remembering to get dressed. I woke him up and told him what had happened. But he merely reminded me of the strict rules concerning heavy drinking at the campus. He's going to take disciplinary action against me. Damned bureaucrats!

"IF WE WERE TO EXPAND THE ELECTRON UP TO BE THE SIZE OF THE SOLAR SYSTEM, THEN WE'VE MEASURED ITS SHAPE ACCURATE TO BE LESS THAN THE WIDTH OF A HUMAN HAIR." WHY IS THIS SURPRISING AND EVEN UNSETTLING FOR SOME PHYSICISTS?

What?!! Of course, physicists are surprised. Less than the width of a human hair?! That's silly!! Personally I am not only surprised, even unsettled, but even outraged! If the electron were so small, what the hell is that ball the size of a marble they've been selling to me as electrons? Do you know how much money I have spent on electrons in the last nine years? Not to mention neutrons and protons!